GERMS AND DISEASE

WHAT ARE PARASITES?

By Kathryn Kroe

Cavendish Square
New York

Published in 2023 by Cavendish Square Publishing, LLC
29 East 21st Street, New York, NY 10010

Website: cavendishsq.com

Library of Congress Cataloging-in-Publication Data

Names: Kroe, Kathryn, author.
Title: What are parasites? / Kathryn Kroe.
Description: New York : Cavendish Square Publishing, [2023] | Series: Germs and disease | Includes index.
Identifiers: LCCN 2021056620 | ISBN 9781502665379 (library binding) | ISBN 9781502665362 (paperback) | ISBN 9781502665386 (ebook)
Subjects: LCSH: Parasites—Juvenile literature. | Parasitology—Juvenile literature.
Classification: LCC QL757 .K76 2023 | DDC 591.7857—dc23/eng/20211124
LC record available at https://lccn.loc.gov/2021056620

Editor: Kathryn Kroe
Copyeditor: Jane Katirgis
Designer: Andrea Davison-Bartolotta

The photographs in this book are used by permission and through the courtesy of: Cover, p. 10 (top left) Kateryna Kon/Shutterstock.com; series art (backgrounds) CHIARI VFX/Shutterstock.com; p. 5 Tatyana Vyc/Shutterstock.com; p. 6 Christoph Burgstedt/Shutterstock.com; p. 9 kokhanchikov/Shutterstock.com; p. 10 (background) Aleksander Bryliaev/Shutterstock.com; p. 10 (top middle) David A Litman/Shutterstock.com; pp. 10 (bottom middle), 11 (top) Rattiya Thongdumhyu/Shutterstock.com; p. 10 (bottom left) ART-ur/Shutterstock.com; p. 10 (top right) wikimedia commons/C_wegeneri.jpg; p. 10 (bottom right) wikimedia commons/Moniliformis_moniliformis.jpg; p. 11 (bottom) KPixMining/Shutterstock.com; p. 12 Joanna Kopera/Shutterstock.com; p. 14 jinnawat tawong/Shutterstock.com; p. 16 D. Kucharski K. Kucharska/Shutterstock.com; p. 17 Irina Bg/Shutterstock.com; p. 19 Designua/Shutterstock.com; p. 21 Roberto Castillo/Shutterstock.com; p. 22 Vera Larina/Shutterstock.com; p. 24 Protasov AN/Shutterstock.com; p. 25 Littlekidmoment/Shutterstock.com; p. 27 vitec/Shutterstock.com; p. 29 Peddalanka Ramesh Babu/Shutterstock.com; p. 31 Sappasit/Shutterstock.com; p. 32 SciePro/Shutterstock.com; p. 34 Andrea Pasquini ph/Shutterstock.com; p. 36 5 second Studio/Shutterstock.com; p. 38 Tuttoo/Shutterstock.com; p. 39 Gorodenkoff/Shutterstock.com.

CPSIA compliance information: Batch #CSCSQ23: For further information contact Cavendish Square Publishing LLC, New York, New York, at 1-877-980-4450.

Printed in the United States of America

CONTENTS

INTRODUCTION

"Parasite" is a term commonly used to describe something or someone that selfishly lives off another. While this description is oversimplified, it is a good place to begin understanding about parasitic organisms. It isn't uncommon to depend on other organisms for survival. Humans, for example, use other living things for food sources. Children depend on their parents for protection. A pet depends on its owner for basic life needs. Depending on others does not make the organism a parasite. If that were true, most living things would fall into this category.

When an organism is parasitic, it does not just depend on another organism. It needs to live in or on that organism, its **host**, to survive. Host organisms do not invite parasites to take up residence; parasites can be quite bothersome. While using another plant or animal, including humans, to thrive, parasitic organisms can take too much. Undetected for too long, they can make the host organism very ill, even causing death.

Parasites can be a threat to humans and many plants and animals that are important parts of our ecosystem. So why is it important to study parasites? How will our understanding of them help us? Being able to identify what exactly parasites are and how they function and make us sick is invaluable in being able to protect ourselves from them.

Dogs depend on their owners to take care of them, but that doesn't make them parasites. Parasites live in or on the bodies of other organisms.

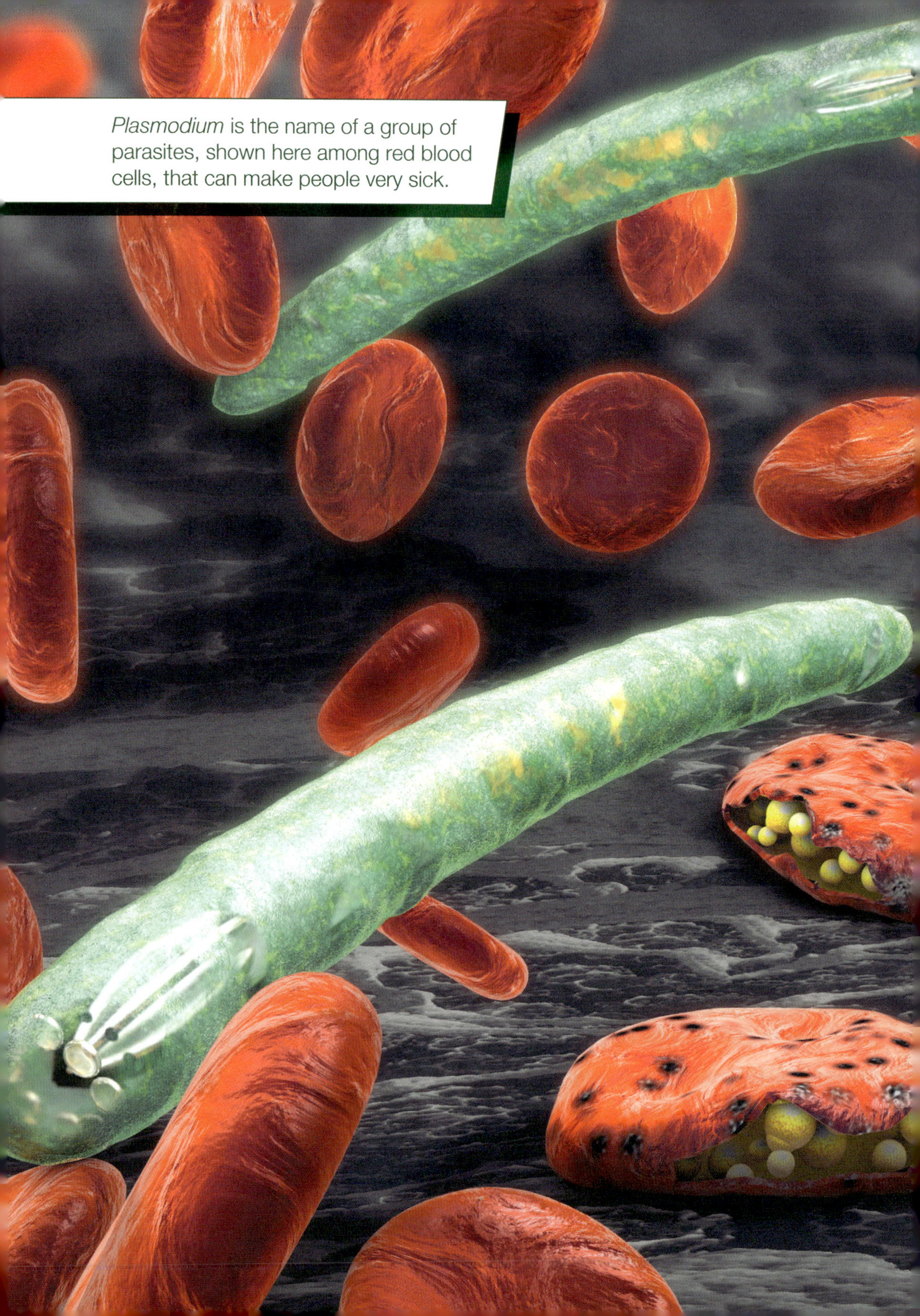

Plasmodium is the name of a group of parasites, shown here among red blood cells, that can make people very sick.

CHAPTER 1

WHAT CLASSIFIES AS A PARASITE?

Parasites set themselves apart by living in or on the bodies of other organisms, or hosts. Without the host, the parasite could not survive. In fact, to be a true parasite, an organism must benefit while harming its host. Usually, the host is much larger than the tiny parasite, as it provides mobility, protection, and a food supply.

While common sense dictates that a parasite would want to keep its host alive, parasites are simple organisms that don't operate with that awareness. If a host becomes too ill because of the parasite feeding off it, the parasite must move to another host or die too.

Parasites are a diverse set of organisms. Some are so small they can be seen only with a microscope, while others can grow to be quite large. Although they are diverse, we can classify them into three groups: protozoa, worms, and parasitic insects and **arachnids**.

PROTOZOA

Protozoa are microscopic, single-celled organisms. These tiny creatures are neither plant nor animal but are classified as somewhere in between. The word "protozoan" is derived from the Greek word for "first," *protos*, referring to the fact that these organisms are some of Earth's simplest life-forms and were probably among the first living creatures to evolve on the planet.

Not all protozoa are parasites. Many, particularly those that thrive in water, do just fine on their own as long as they are in the right environment. Amoebas, for example, get their food and oxygen from the water in which they live. Some protozoa are green because, like plants, they have organelles containing chlorophyll that produce food from sunlight. Still other protozoa do need a host in order to survive, making them parasites.

Protozoa can be divided into four categories: flagellates, sarcodines, apicomplexans, and ciliates. Flagellates possess hairlike appendages that enable them to move quickly and smoothly through water. Many are harmless, but some, like the group called the trypanosomes, can cause disease.

Sarcodines use pseudopods, or "false feet," to move. The pseudopod resembles a tiny toe. Some sarcodines can cause disease when ingested, but many are harmless. One ocean group grows a skeleton that is shed. Over time, the skeletons build up and create structures similar to shells other animals use for shelter.

Unlike flagellates and sarcodines, all apicomplexans are parasites. They move by gliding through their environment. A number of apicomplexans can be quite dangerous to humans. **Malaria**, for example, is caused by an apicomplexan parasite.

Ciliates are like flagellates in that they also possess hairy growths on the exterior of their cells to help them move. More developed in structure, however, ciliates are the most complex kind of protozoan. They may also have other complex anatomical features to help them survive, such as tubes that aid in feeding.

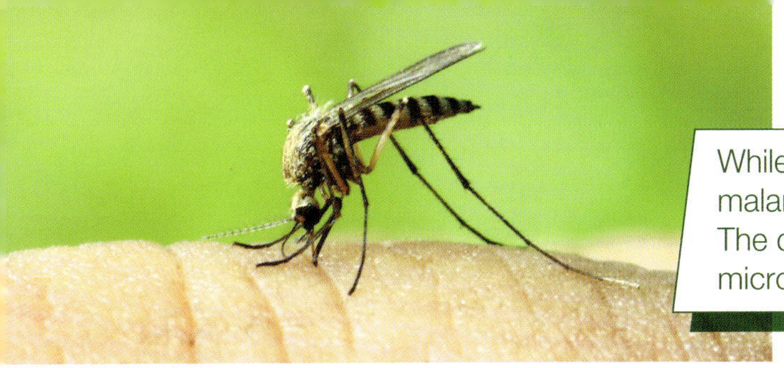

While mosquitoes are known to carry malaria, the insect is not the parasite. The disease is actually caused by a microscopic protozoan parasite.

Helminths

Worms are animals that possess a soft, thin body without a backbone or limbs. You have likely seen worms in your backyard or at a park. That is because worms often live in soil or water. Many are essential to their environments, serving as cleanup crews and adding beneficial nutrients to the soil. Parasitic worms, also known as helminths, on the other hand, can pose a danger to plants, humans, and animals. Parasitic worms feed off their host, often without the host knowing the worm is present. This can lead to blood and nutrient loss, causing health disorders.

Parasitic worms are classified into three primary groups: flatworms, roundworms, and thorny-headed worms.

Flatworms indeed are flat but come in different shapes. Some are narrow, while others grow in an oval form. Flatworms and other worms that do not live as parasites are free to move about in the soil or water and have a variety of food sources. However, parasitic flatworms, such as flukes and tapeworms, rely on other creatures for their nourishment. They can pose serious health problems to humans and other animals.

Roundworms possess long, tubular bodies. This is the largest group of worms, with approximately 12,000 species. Most free-living species inhabit the water and soil and are perfectly harmless. Parasitic roundworms, however, can cause disease.

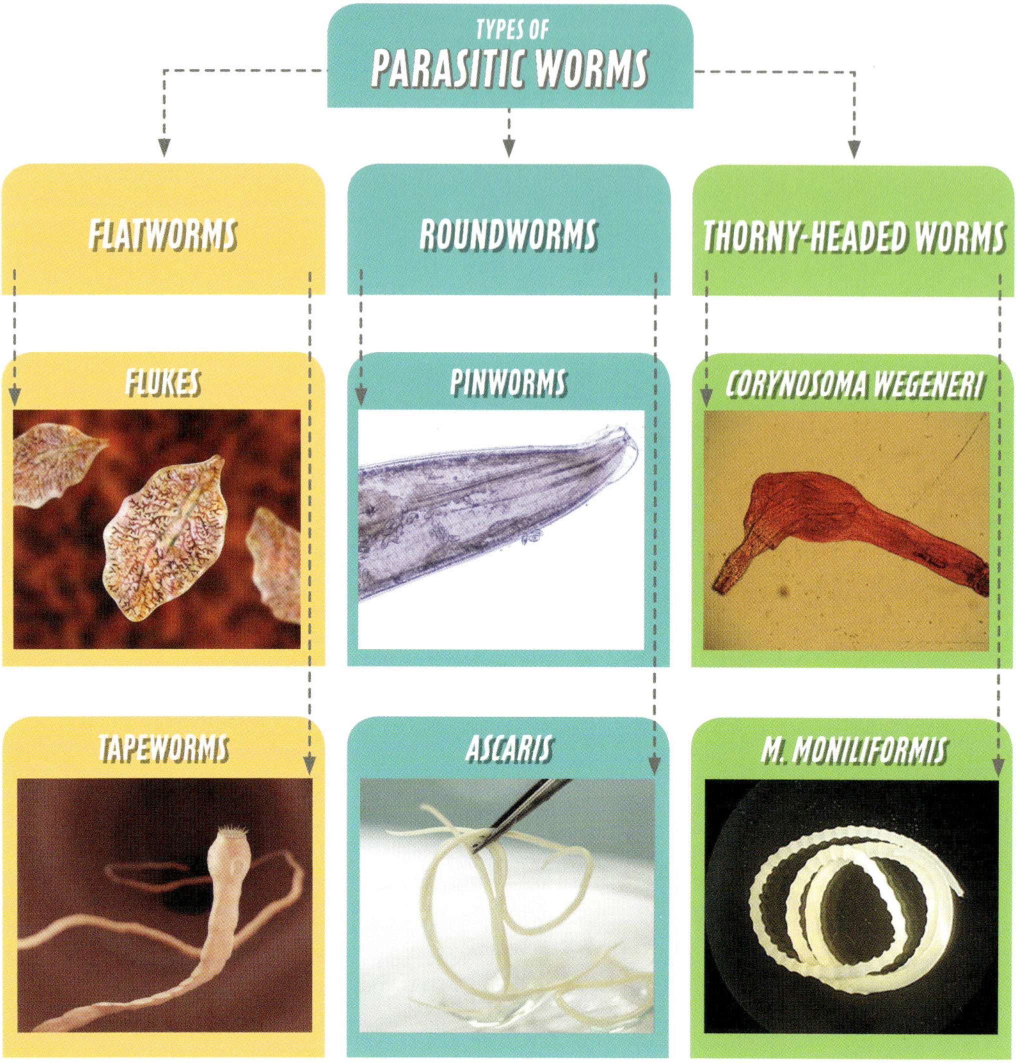

Parasitic worms can be classified into three main categories. In the graphic organizer above, you can see examples of each type of worm.

Thorny-headed worms, as the name implies, are characterized by a hook-shaped snout. They use this to attach themselves to the **gastrointestinal** (GI) tract of a host to feed.

While it is sometimes disputed, some scientists consider a segmented worm such as a leech a parasite too.

This type of worm, called a thorny-headed worm, uses its hook-like snout to attach itself inside a host.

Parasitic Insects and Arachnids

Insects are small, air-breathing animals that are invertebrates, meaning they do not have a backbone. They are usually characterized by their six legs and bodies divided into three parts. They also have an exoskeleton, a hard shell covering their soft body parts. They may or may not have wings. Arachnids are a type of animal with eight jointed legs and segmented bodies divided into two parts. Spiders are one common arachnid, but the group also includes ticks and mites, which are parasitic.

Except for one species of mite that lays its eggs in human skin, insect parasites and arachnids live on the skin instead of inside the bodies of their hosts. Fleas and lice are common parasitic insects. Ticks are common parasitic arachnids. They have mouthparts that

Do you see the eight legs and segmented body on this tick? These features make the parasite an arachnid.

PERFECT PAIR?

If you've ever seen the animated film *Finding Nemo*, you might remember that Nemo and his dad, who are clownfish, live in a sea anemone. This is actually an example of the symbiotic relationship called mutualism that occurs in nature. Both organisms are able to survive better with the other. A sea anemone provides shelter and protection for the clownfish, while the clownfish provide nutrients through waste and lure in other fish the sea anemone can eat.

Symbiosis is defined as a close relationship between individuals of two or more different species. Three symbiotic relationships we often hear about are mutualism, commensalism, and parasitism. Mutualism, like the clownfish and sea anemone, describes a partnership where both species benefit. Commensalism is a relationship where one species greatly benefits and the other is not harmed or affected. Parasitism describes a symbiotic relationship where one species benefits and the other species is harmed. Sometimes all species that participate in symbiotic relationships are mistakenly considered parasites, but the difference in classification is important when understanding truly parasitic organisms.

The relationship between these two organisms is not parasitic, but it is an example of symbiosis.

enable them to attach themselves to the skin, where they can draw out nutrients.

While mites, fleas, ticks, and lice tend to target one species as hosts, another group of insects prefers to take a small amount of blood from a number of different hosts. They are known as blood-sucking micropredators. Mosquitoes and biting flies such as tsetse flies, sand flies, and black flies are micropredators. Technically, they are not parasites, but they can carry protozoan parasites in their bodies and spread disease.

Many insects live free for part of their lives and are parasitic at other times. For example, fleas begin life as eggs. When hatched, the larvae are not parasites. When the fleas become adults, then they become parasites.

When identifying whether an organism is a parasite, ask yourself whether the organism benefits by harming another. This is an important distinction when understanding parasites. While we classify parasites into only three main categories, the diversity among the organisms that fit into them is great.

WHAT DO YOU THINK?

1. Can you name examples of the three types of symbiosis?
2. If worms can grow to be quite large, how do you think they find an unknowing host?
3. Why do some people think mosquitoes and biting flies are parasites?

Just because water appears clear does not automatically make it safe to drink. If water is not filtered properly, microscopic parasites can travel in the water to their new hosts.

CHAPTER 2

HOW DO MICROSCOPIC PARASITES SPREAD?

Germs make human beings sick. Viruses, bacteria, and fungi are classes of germs, besides protozoa, that are responsible for illness and disease. Parasitic protozoa that cause disease are called pathogenic. A pathogenic protozoan benefits from its host, often making it seriously ill. Protozoa are different from other germs; they have the ability to move on their own. Once they infect, they don't need cells to spread.

The most common way to become infected with parasitic protozoa is by ingesting them. These pathogens are microscopic, so even if food or water does not look dirty, it can be **contaminated**. Once someone eats or drinks contaminated food or water, the parasites in them are able to make them their new host.

WATER PARASITES

Waterborne disease is still fairly common in the United States, with many states reporting illnesses each year. Two of the most common diseases caused by protozoa in water are giardiasis and cryptosporidiosis. Both get their names from the parasitic protozoa that cause the illnesses.

Giardiasis occurs when a person or animal becomes infected with a parasite called *Giardia lamblia*. This parasite can be found in soil, food, and water, and on moist, contaminated surfaces. Ingesting water containing the parasite is a common way in which giardiasis is spread. Frequently, water will contain sewage or feces that harbor the parasites. *Giardia lamblia* is especially hardy and can survive in a dormant state for extended periods of time. That is because the parasite is covered with a protective outer shell. It is impossible to detect in water unless viewed under a microscope.

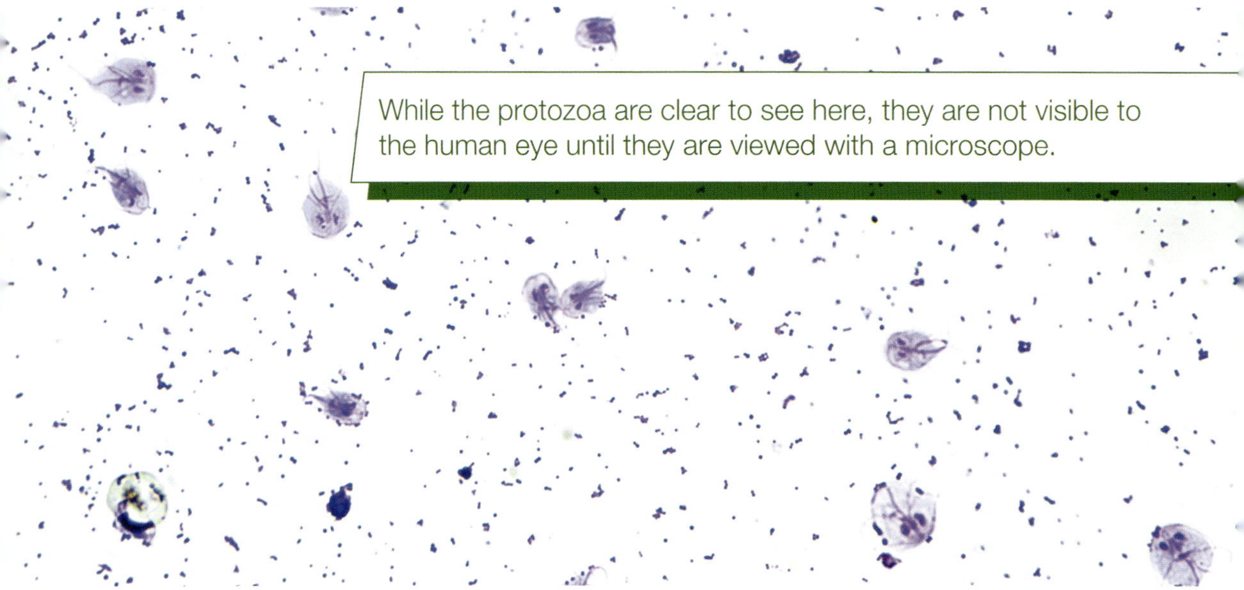

While the protozoa are clear to see here, they are not visible to the human eye until they are viewed with a microscope.

Symptoms of giardiasis include diarrhea and other intestinal problems, stomach cramps, and sometimes nausea, dehydration, and weight loss. Symptoms usually last anywhere from two to six weeks but may last longer. Sometimes outbreaks occur after flooding, when sewage or waste runoff enters the water supply. A person can also become infected after swallowing contaminated water found in improperly treated swimming pools, hot tubs, or fountains, or in natural settings such as lakes, rivers, springs, ponds, and streams. Giardiasis is usually treatable with prescription drugs.

A tiny parasite called *Cryptosporidium parvum* causes the diarrheal disease cryptosporidiosis, better known as crypto. Similar to the giardiasis parasite, the crypto parasite possesses a shell that enables it to survive outside of its host or in certain environments for long periods.

Some parasites have similar symptoms to viruses that affect the gastrointestinal system, but the germs causing them and the treatment are quite different.

Protozoa in Food

Food, like water, can provide an environment in which parasites can live and wait for a host, such as an unsuspecting hungry human. Two parasitic diseases commonly spread through contaminated food, in addition to water, are amebiasis and cyclospora.

The single-celled parasite called *Entamoeba histolytica* causes the intestinal illness amebiasis. It is a common disease in countries with poor sanitary conditions. The symptoms of amebiasis are usually mild, but the disease can progress to a severe form called amebic dysentery. With amebic dysentery, the victim may experience severe stomach pain, fever, and even damage to the lungs or brain. The parasite literally invades the body, where it grows and spreads. If left untreated, amebic dysentery can be fatal.

People become infected with the parasite after swallowing contaminated food or water or by touching cysts, which are the eggs of the parasite, and then accidentally bringing the eggs to the mouth. The cysts, invisible to the naked eye, could be present on surfaces contaminated with the parasite.

Cyclospora, as with most diseases caused by parasitic protozoa, is spread by drinking contaminated water, by eating contaminated food, or by hand-to-mouth contact after touching a surface with the parasite on it. It takes days, even weeks, for a person to feel the first symptoms of an infection by cyclospora, which causes intestinal problems. That is because the parasite multiplies in the body after passing through the liver. If left untreated, victims of the disease could suffer for a month or longer, during which time there may be periods of feeling better followed by relapses.

Insects as Vectors

Insects and other animals that can spread disease are known as vectors. Vectors are living organisms that transmit an infectious agent from one organism to another. In virtually all cases, the insects themselves are not harmful unless present in large numbers. They are dangerous to animals and humans because of the parasites, bacteria, and other germs that they might carry and then transfer to bite victims.

Mosquitoes and other flying, biting insects are one group of insect vectors. These insects usually possess a mouthpart similar to a needle. They puncture the skin of victims and draw out blood. If the insect has parasites in its body, the parasite might spread into the victim's blood or lymphatic system during the bite. Diseases that might not otherwise travel easily from person to person can spread rapidly in the presence of biting insects. A mosquito, for example, might bite an infected person. That mosquito could then bite another person, and then another, spreading the parasitic disease. Worldwide, malaria is one of the most serious and widespread parasitic diseases. It is spread by mosquitoes. It is said that more than 40 percent of the world's population is at risk. In the United States, most cases occur among immigrants and those who travel to areas where malaria is a concern.

MALARIA LIFE CYCLE

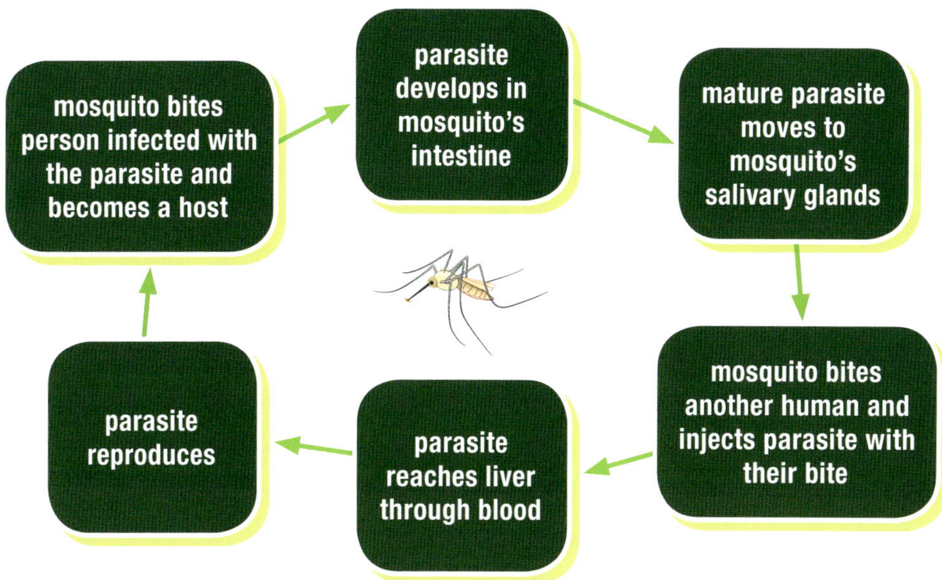

The parasite that causes malaria doesn't live in just one host, which extends its life cycle. In the flow chart above, you can see how the parasite moves from mosquito to human to keep reproducing.

Fleas and ticks have been among the most common vectors for various viral and bacterial infections. Many of the **insidious** diseases known throughout human history to be spread by these creatures have been largely eliminated in the modern era or now have more effective treatments. The bacterium *Yersina pestis*, for example, was transmitted by fleas (as well as rodents with fleas on them) and was responsible for the devastating plague pandemic known as the Black Death, which may have killed as many as 200 million people

PARASITES ON OUR PETS

If you have a house pet, you may be aware of fleas and ticks being a common bother. These parasites love dogs and cats as their hosts. We need to treat our pets with medicine to prevent these parasites from getting comfortable. Flea and tick preventatives and treatments work in different ways, but the goal is always to prevent a parasite from feeding on the animal and to kill off any existing parasites and their eggs.

Besides keeping your pet comfortable, keeping fleas and ticks away keeps human occupants of the house safe from them too. While fleas don't usually find humans to be hospitable hosts, they can still bite to feed before returning to a pet and pose a risk as a vector for microscopic parasites. Ticks can also transfer from their pet host to a human—and often will.

If you find a tick using you as a host, it is important to remove it carefully because viruses, bacteria, and parasites can be present in the tick's body. This should be done with a pair of fine-tipped tweezers. Grasp the tick as close to the skin as you can, and lift it out, being careful not to squeeze the tick's round body. It is a myth that lighting a match or using petroleum jelly will help. They can even do more harm because if the tick becomes irritated, it may tighten its grip and inject even more potentially dangerous fluid into the host.

in 14th-century Europe. Today, plague can be treated with drugs.

Microscopic parasites spread in a variety of ways that are harder to prevent because we cannot see them. Still, understanding where and how they travel gives us important information as to how to keep our bodies safe from becoming hosts.

The Black Death, an outbreak of bubonic plague, was devastating when it spread in the 14th century. Fleas and the rodents that transported them were responsible for the rapid spread in crowded towns. While fleas are parasites themselves, the pathogen they were a vector for was a bacterium, not a parasite, in this case.

WHAT DO YOU THINK?

1. If we can't see them, how can we prevent the spread of microscopic parasites?
2. What makes it hard to protect against malaria?
3. Why are fleas and ticks parasites and mosquitoes are not?

While fleas are small, these hopping parasites are certainly visible, especially when many are using the same host.

CHAPTER 3

HOW DO LARGER PARASITES SPREAD?

Not all parasites are microscopic; some are large enough to see with the naked eye. As mentioned earlier, even though they are often vectors, fleas and ticks are also parasites themselves. As adults, these wingless creatures must live on other creatures such as birds and mammals, including humans, in order to survive. Unlike most mosquitoes, which are another parasite vector, fleas and ticks can pose a health risk even if they are not infected with a parasitic germ.

Lice and certain types of worms are other common parasites that are usually visible, especially in their adult stages.

Lice

The parasitic insect known as a louse is so common that you probably know someone who has suffered from an **infestation**. Every year, 6 million to 12 million people contract head lice, a species of lice adapted to living on the human scalp and neck.

Lice spread only when there is direct contact with a person who has lice or with items that touched the affected person's hair. These items may include clothing, hats, combs, carpets, beds, pillows, couches, or bath towels. Often, contact occurs during recess, sports, slumber parties, and camping trips.

Lice are common parasites. While not particularly harmful to our health, they can make people quite uncomfortable.

When lice move into someone's hair, they can go through their entire life cycle right on that person's head. Lice begin as nits, or eggs. The eggs look like yellow or white specks and can be confused with dandruff, or dried skin particles. The eggs attach to the hair shaft where they later hatch.

Symptoms of lice infestation include itching, a tickling feeling on the head, and visible signs of the parasites themselves. While relatively harmless, lice are hardly welcome guests. They can lead to infections if scratching produces sores. The good news is that they are highly preventable. To avoid infestation, take care not to share combs, brushes, or hats with others. Take particular care not to come into contact with the head of a person you know is suffering from head lice. If you go on a sleepover, bring your own pillow, pillowcase, and sleeping bag or sheets.

Lice can cause the head to become very itchy when they attach themselves to feed. Luckily, once lice are detected, it is usually easy to rid the scalp of the infestation.

THE STIGMA OF LICE

Head lice are very common parasites, but because creepy crawly things are unwelcome on our bodies, an infestation holds a negative **stigma** for the unwilling host. Some people may think that a person must be dirty for bugs to want to make their home on their head, but this is a myth. In fact, lice are just as likely, if not more, to move in on a clean scalp as a dirty one. Hygiene is not a concern to these crawling pests. Lice themselves aren't known to spread any disease either. We may be grossed out by them, but overall, they are just a nuisance and not dangerous.

Another common myth that has contributed to this stigma over the years is that lice spread easily in schools. Because of this idea, classmates may have avoided someone suspected to have lice. This specific spread has not been proven true. Lice can only spread from direct contact, as they don't hop or fly but only crawl. It is much more common for lice to spread from things such as pillows, hairbrushes, and sheets that are not usually used in schools.

If you do contract head lice, they are highly treatable with medicated shampoos, creams, and ointments. Usually a single application of medicine works. In some cases, lice can spread to the eyebrows and eyelashes, which may require special treatment by a doctor. Other types of lice can affect different parts of the body.

WORMS

Worms can be more dangerous to humans than lice because worms may live and multiply inside a person's body, unlike lice, which exclusively live on the skin's surface. It is important to remind

While you may not like their wriggly, slimy nature, the earthworm is not a worm to fear. This worm is not parasitic and has an important role in its ecosystem.

yourself that a parasitic worm is very different from most worms that you find in the garden. Common garden worms are usually beneficial to humans because they help break down soil and keep it nutritious for crops and other plant life.

While soil worms feed on the organic matter in dirt, parasitic worms feed off whatever they infest. There are many different kinds of parasitic worms. Virtually all begin by feeding on food and other nutrients stored within the host's body. If this supply dwindles as the worms produce and multiply, the parasites may begin to feed on blood and other body fluids. When this happens, the host may suffer from anemia (a lack of red blood cells), infections, and other health problems.

There are four principal methods of transmission, or ways in which people can become infected with parasitic worms. Worms spread to humans through water, food, insect bites, and contaminated soil. Most parasitic worms that live in water can also live in food. That is because particles of food or waste material may wind up in water, creating a breeding ground for parasitic worms. Municipal treatment facilities carefully filter water and use certain chemicals, such as chlorine, to rid water of parasites and other contaminants. Contracting worms from water is not common in the United States and other developed countries. Use of improperly tested well water, however, can lead to parasitic diseases. Consuming water after a flood can as well. Swimming in contaminated water is another way in which parasites can spread to humans. Some worms spread when a person eats contaminated food or accidentally swallows parasitic worms after touching a contaminated surface. These are the most common ways to get infected, particularly in the United States.

Schistosomiasis

In certain parts of Africa, South America, the Caribbean, China, the Middle East, and Southeast Asia, some types of snails harbor parasitic worms called schistosoma, or blood flukes, that cause a disease known as schistosomiasis. The parasites can penetrate the skin of a person who may be swimming or bathing in water where the snails live, and then enter the person's bloodstream.

Mild infections cause rash and flulike symptoms. In more severe cases, the parasites can travel to the brain and spinal cord. Such an infection can even cause paralysis. Other major organs may also be damaged. Thankfully there are prescription drugs that can kill the worms and clear up the infestation.

This parasite is a schistosoma. It can cause severe damage to a person but can be killed easily with prescription drugs.

Pinworms

One of the most prevalent infection-causing parasitic worms is the pinworm. This is a small, white worm that resides in the human intestines. A sneaky worm, the pinworm leaves the intestines at night. Under the cover of darkness, females exit the body and lay eggs on the sleeping person's skin.

Young children are most often affected, as the worms often spread in child care centers, day camps, schools, and other places where close contact with victims might occur. Eggs laid on the skin can fall off and live for up to two weeks on clothes, bedding, and other objects. Usually a person becomes a host after accidentally swallowing one or more of the microscopic eggs.

The Ascarid Worm

The ascarid worm is similar to the pinworm and is by far the most common parasitic worm that infects humans worldwide. In the United States, the presence of ascarids, small intestinal worms, usually occurs in rural areas of the Southeast. Among livestock, the animal most affected by ascarids and other parasitic worms is the pig. That is because pigs eat close to the ground, may be fed contaminated food scraps, and spend a lot of time rolling around in soil where the worms might reside.

Tapeworms and Roundworms

If humans eat contaminated meat that has not been properly cooked, they can become infected with parasitic worms. Two types of worms spread by eating contaminated pork and beef are tapeworms and roundworms.

Tapeworms, which live in the small intestine of human hosts, can grow and reproduce for many years if left untreated. Some tapeworms may reach several feet in length. A single worm also can lay 50 eggs a day.

Trichinosis, caused by a roundworm, occurs when a person eats raw or undercooked pork infested by the parasitic worm. It causes everything from fever to heart problems. Trichinosis used to be common in the United States, but strict regulations on the way pigs are raised and butchered, and better recommendations for cooking pork, have nearly eliminated the problem.

Tapeworms are a type of flatworm that can grow very big. Untreated, they can cause many health issues for their host.

WORMS IN SOIL

One of the creepiest-looking intestinal parasites is the hookworm. The tip of a hookworm is a head with a giant open mouth full of sharp teethlike structures. They allow it to hook on to the walls of a human being's intestines, where it feeds. The eggs of the hookworm are passed through human feces or that of any animal infected with them. People can get the worm by walking barefoot in contaminated soil, or, in the case of certain types, by accidentally swallowing the larval form of the worm. Those serving as a host for the worm for the first time are most likely to have gastrointestinal problems, but many who carry hookworms exhibit no symptoms.

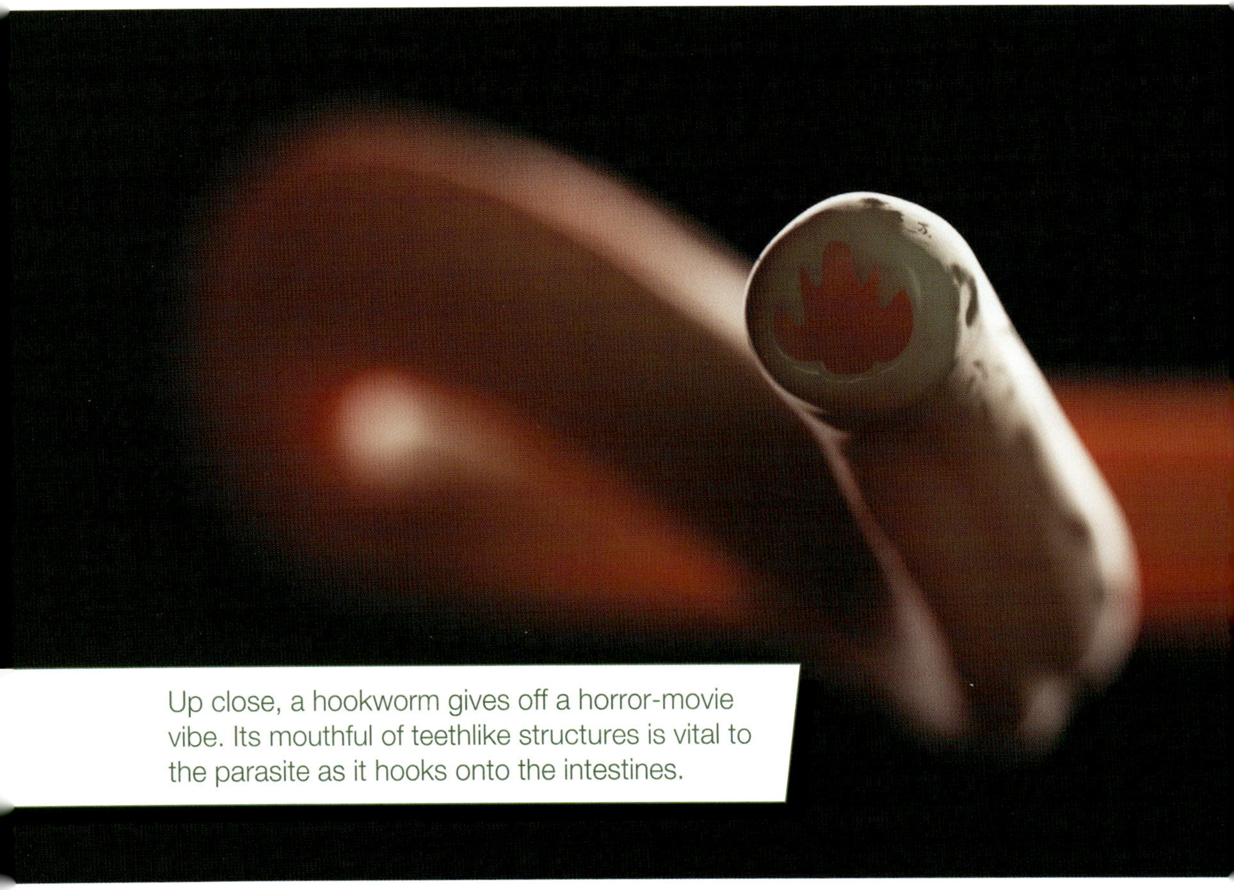

Up close, a hookworm gives off a horror-movie vibe. Its mouthful of teethlike structures is vital to the parasite as it hooks onto the intestines.

We may think larger parasites would have a hard time finding unsuspecting hosts, but most of the parasites that fit into this category are still quite small. Though they are not microscopic, they are still able to thrive, often without being detected. Some of these parasites can even grow larger once they find a host organism. Identifying common methods of transmission is the best way to keep ourselves from playing host to these parasitic worms and insects.

WHAT DO YOU THINK?

1. Why do you think people believe lice spreads easily in schools even though it has not been proven?
2. How do you think we can prevent worms from living in food we eat?
3. Why can't our first line of defense, the skin, always protect us from parasites?

While tap water in most of our homes is treated to be safe to drink, that is not always the case. When you visit a new place, you should always ask whether the water is safe to drink straight from the tap, and use an alternative if it's not.

CHAPTER 4

HOW CAN WE PREVENT PARASITES?

While many parasites can be treated with prescription drugs, some treatments are expensive, rare, or have unwanted side effects. Instead of relying on medicines, preventing parasites from affecting us is the best method to keep safe. To do this, we need to understand how parasites spread.

Prevention of Foodborne and Waterborne Parasites

Parasites can pass to humans through water, food, and soil. An infected mosquito or other insect can also pass on a parasite through its bite. While there are many kinds of parasites that multiply quickly, it is possible to stop the cycle from reaching humans.

Tap water in the United States must meet certain standards, so infection from municipal water is rare. If you are drinking untreated water, perhaps during a hike or camping trip or if you are traveling to countries without proper water treatment facilities, water still can be made safe to drink.

Boiling water for a minute or more usually kills any parasites that might be present. **Iodine** tablets, available at most pharmacies and sporting goods stores, also can kill parasites when the tablets are dissolved in the water. Filters containing iodine can be used to protect against protozoa and worms.

Boiling water is an effective way to kill parasites and most other germs too. If you are unsure if water is safe to drink, boiling is always recommended.

When traveling outside of the United States to countries with known water problems, it is best to drink canned or bottled beverages. Avoid ice, as many parasites can survive below the freezing point of water. Boiled water drinks such as tea are safer.

Be sure to thoroughly cook foods, as instructed on the packaging or in a cookbook. Fish, for example, requires less cooking time than pork, which must be heated to a certain internal temperature. When storing foods for transport, keep hot foods hot and cold foods cold using a thermos or some other special, protective container.

When eating in countries with poor **sanitation**, avoid salads, other raw vegetable and fruit dishes, and **unpasteurized** milks and cheeses. Salads, for example, may have been washed with contaminated water. Also take care when purchasing food from street vendors. Make sure the food is fresh and heated before eating.

Prevention of Parasitic Insect Illnesses

The best way to prevent being bitten by a mosquito, flea, tick, or other animal associated with parasites is to avoid the bugs altogether. Most disease-spreading mosquitoes are active at dawn or dusk, when temperatures are not too warm or cold. You may have even noticed that more mosquitoes fly around outside your home during these times. It is then that they are likely to feed on unsuspecting hosts such as humans.

Fleas and ticks are a bit harder to track down. They can exist in certain rural areas, in tall brush or grass, or even in the fur of a family pet. Since it is impossible to completely avoid mosquitoes and other insects, sometimes other precautionary measures are necessary. Be sure to regularly wash the bedding of dogs and cats. Keep the animals well brushed and groomed. Bathe them when necessary. Cats do not always require baths, so check with your veterinarian first to ask about your specific pet.

When in outdoor areas where mosquitoes and parasitic insects are prevalent, wear protective clothing. Bugs cannot bite through most clothing, so anything that is covered will most likely receive some protection. In rural places where there are a lot of fleas and ticks, pant legs can be tucked into socks, which will prevent fleas and ticks from touching your skin.

A number of insect repellents are available, even in grocery stores. Use them with the guidance of adults and only when protection from clothing and other methods are not enough.

In areas where malaria spreads, mosquitoes are even a threat during the night. Mosquito nets placed around beds protect people while they are sleeping.

MODERN SCIENCE

Modern research, medicines, and other tools in the war against parasitic diseases help to keep parasites under control. Current problems, such as pollution, overpopulation, **bioterrorism**, and climate change, however, often work in favor of parasites. The

Parasites are diverse organisms, and understanding them further can only be helpful to humanity.

quality of water, food, and soil all suffer when under stresses that upset the natural balance. A fish weakened from bacteria present in a polluted lake, for example, would be more vulnerable to infestation by parasites. It is therefore important to address problems that affect ecosystems.

While parasites are rightfully considered a serious threat to human health throughout the world, scientists believe they and their

PARASITES AS MEDICINE

While modern science continues to investigate how some parasites could potentially be beneficial, one parasite has been used in medicine since the time of ancient Egypt. Leeches are bloodsucking parasites that have been used to treat all sorts of problems and infections, both successfully and unsuccessfully, throughout history.

In the 19th century, bloodletting made a comeback as a popular treatment in medicine. It involved draining the blood, often with leeches, to reduce blood flow and, in theory, treat disease. In general, this practice did not work, and as medical professionals gained more understanding of germs and disease, the practice was largely abandoned.

Today, leeches are actually an approved treatment for specific instances after reconstructive plastic surgery. They help carry blood away from surgical areas of the body. In this instance, leeches can prevent tissue death. Besides moving stagnant blood, leech saliva contains an anticoagulant that can prevent blood clots. While many patients and families are hesitant to accept the treatment due to their slimy nature, leeches can be lifesaving in some cases.

life cycles can provide lessons that will improve medical knowledge. Researchers are investigating how tapeworms and other parasites survive in adverse environments—including the inhospitable surroundings of the human digestive system—and hope to use this information to improve **pharmaceuticals**, treat different kinds of illnesses, and cure parasitic infections themselves. One aim is to design drugs that can withstand extreme conditions using the defense mechanisms and physical makeup of parasites as an example. If these projects yield results, parasites may inadvertently contribute to helping humanity.

Like other germs and diseases, those spread by parasites have been around since the beginning of time and will certainly continue to be a threat in the future. The better we understand and treat known parasites, the further prepared we can be for health problems caused by this group of organisms in the future.

WHAT DO YOU THINK?

1. Do you think clean, safe drinking water should be considered a basic human right? Why or why not?
2. Do you and your family follow any of the methods in this chapter to keep your food and water safe from parasites and other germs? If you do, which ones do you practice?
3. How might taking better care of the environment help stop the spread of dangerous parasites?

GLOSSARY

arachnid: An animal posessing eight legs and a body divided into two parts.

bioterrorism: The intentional release of germs that can sicken or kill people, livestock, or crops.

contaminate: To soil, stain, or infect by contact or association.

gastrointestinal: Relating to the stomach or intestines.

host: The organism that a parasite lives in or on.

infestation: The presence of an unusually large number of insects or animals in one place, typically causing damage or disease.

insidious: Proceeding in a gradual, subtle way but with harmful effects.

iodine: A chemical element often used with alcohol as a mild antiseptic.

malaria: A serious and widespread disease spread by mosquitoes carrying parasitic protozoa that can cause fever, body aches, fatigue, and even death.

pharmaceutical: A substance manufactured for use as a medicinal drug.

sanitation: Conditions relating to public health, especially clean drinking water and adequate sewage disposal.

stigma: A mark of disgrace associated with a circumstance, quality, or person.

symbiosis: An interaction between two different organisms in a close physical association.

unpasteurized: Relating to substances not heated to high temperatures for a short time in order to kill bacteria or other microorganisms present.

FIND OUT MORE

BOOKS

Lundgren, Julie K. *Gross and Disgusting Parasites*. New York, NY: Crabtree Publishing Company, 2022.

McInerney, Joanna. *Made for Each Other*. London, UK: Big Picture Press, 2021.

Tilden, Thomasine E. Lewis. *Dangerous Worms: Parasites Plague a Village*. Danbury, CT: Children's Press, 2020.

WEBSITES

CDC: Parasites
cdc.gov/parasites
This website provides a wealth of information on parasites, including information for specific groups of people such as travelers, women, children, and more.

Peekaboo Kidz
youtube.com/c/PeekabooKids
This YouTube educational entertainment channel has videos on different scientific subjects, including a general overview on parasites and more specific videos for certain parasitic organisms.

Smithsonian National Museum of Natural History
naturalhistory.si.edu
This website has educational material on a multitude of subjects, such as parasites and parasitism, and includes teaching and learning resources.

ORGANIZATIONS

The American Society of Parasitologists (ASP)
President Matt Bolek
Oklahoma State University
Department of Zoology
501 Life Sciences West
Stillwater, OK 74078
amsocparasit.org
The American Society of Parasitologists (ASP) is a diverse group of about 700 scientists who are interested in the study and teaching of parasitology.

Centers for Disease Control and Prevention (CDC)
1600 Clifton Road
Atlanta, GA 30329
www.cdc.gov
The Centers for Disease Control and Prevention (CDC) is the primary U.S. government agency in charge of tracking, treating, and preventing public health threats, including infectious diseases.

World Health Organization (WHO)
Avenue Appia 20
1211 Geneva
Switzerland
www.who.int/en
The World Health Organization (WHO) is the specialized agency of the United Nations primarily concerned with international public health.

Publisher's note to educators and parents: Our editors have carefully reviewed these websites to ensure that they are suitable for students. Many websites change frequently, however, and we cannot guarantee that a site's future contents will continue to meet our high standards of quality and educational value. Be advised that students should be closely supervised whenever they access the internet.

INDEX

About the Author

Kathryn Kroe has a master's degree in public health from the University at Buffalo and a bachelor's degree in education from SUNY Buffalo State College. She is passionate about spreading accurate information and helping young people understand issues pertaining to their health and the health of those around them. She currently lives in Lockport, New York, with her husband and two toddlers. She enjoys walking with her family, playing with her Australian shepherd, Zoey, and traveling to see new places.